Y0-CQV-888

STRANGE PLANTS

THE EARTH'S GARDEN

Jason Cooper

Rourke Enterprises, Inc.
Vero Beach, Florida 32964

PHOTO CREDITS

All Photographs © Lynn M. Stone

LIBRARY OF CONGRESS
Library of Congress Cataloging-in-Publication Data
Cooper, Jason, 1942-
Strange plants / by Jason Cooper.
p. cm. — (The Earth's garden)
Includes index.
Summary: Describes a variety of unusual plants and plantlike forms, including air-plants, carnivorous plants, lichen, etc.
ISBN 0-86592-625-5
1. Plants—Juvenile literature. [1. Plants.]
I. Title. II. Series: Cooper, Jason, 1942- Earth's garden.
QK49.C72 1991
581—dc20 91-7140
CIP
AC

Printed in the USA

TABLE OF CONTENTS

STRANGE AND CURIOUS PLANTS

We all know that flowers, trees, and grass are plants. They have green leaves and grow from roots in the ground.

Some plants, however, are different from the common plants we know best. We may not even know they are plants.

Some of these unusual plants have a taste for insects. Some do not grow in the ground. Others have no green color at all.

Epiphytes covering tree branches in Florida forest

PLANTS THAT PERCH ON BRANCHES

Epiphytes, or air plants, are plants that usually live in trees instead of in the ground. Their roots attach themselves to tree bark.

Epiphytes live on rainwater, sunshine, and whatever food their roots can soak up.

Air plants include some ferns, orchids, and plants called **bromeliads.**

Many of the air plants are built in the shape of cups so they can catch rainwater. The water in them attracts snails, insects, and treefrogs.

Ferns, air plants, and orchids growing in jungle

SPANISH MOSS

One of the best-known air plants is Spanish moss. Spanish moss grows in the warm climate of the southern United States. It is neither Spanish nor real moss.

Spanish moss hangs from branches in great, silvery beards. Like other air plants, Spanish moss lives on sunshine and rain. Its stems and leaves are covered by tiny hairs. These hairs soak up water for the plant's use.

Spanish moss on an oak tree

SEAWEED AND OTHER ALGAE

Algae (pronounced AL jee) are plants of many forms and colors. They live in many places, from the cold Arctic to the hot tropics.

Seaweed is a common alga (AL guh). The giant kelp is a ribbonlike seaweed alga. It can be over 150 feet long.

Much of the green matter that grows on freshwater ponds is also an alga.

Algae live in freshwater springs where the water is too hot to touch. They also live on the backs of turtles and in the snow of mountains.

California sea otter wrapped in kelp

Banyan trees in Florida

White ibis standing on the roots of red mangrove trees

PAINTED ROCKS AND TREES

One of the most curious living things is lichen (LIE ken). Lichen is a mixture of a plant—algae—and fungus. Funguses are plantlike types of life, but they are not plants. (Mushrooms are a well-known kind of fungus.)

Lichen grows on rocks, trees, and soil. People often mistake colorful red, orange, and white lichens for paint!

Lichens may be smooth, crusty, threadlike, or lacy. A lacy ground lichen known as reindeer moss is a favorite food of reindeer in the Far North.

Rocks "painted" with lichen in Manitoba, Canada

PLANTS THAT TRAP

Among the strangest plants are insect-eating plants. These plants have bright flowers and green leaves. But otherwise, there is nothing ordinary about them.

Insect-eating plants trap insects. The plants use part of the insects they catch as food.

Insect-eating plants trap insects in their leaves. Some leaves of insect-eating plants are coated with sticky liquids. Others form a trap door or a deep pitcher that traps insects.

Mosquito trapped by insect-eating sundew plant in Indiana

CANDLES IN THE FOREST

The snow plant blooms each spring in the Sierra Mountains of California. Unlike most plants, it has no green color. It stands out like a bright red candle.

Another curious plant of forests is the waxy, white Indian pipe. It is named for its shape. Like the snow plant, Indian pipe has no green color.

Indian pipe and snow plants lack **chlorophyll.** Chlorophyll is the green substance that most plants use in making food.

Indian pipe in Connecticut forest

REMARKABLE TREES

Each **species,** or kind, of tree is different from the others. But some trees are especially different. The bald cypress, for example, surrounds itself with wood growths called knees.

The red mangrove tree grows on tall, curved roots that lift it above the ocean water in which it lives.

Banyan trees also have roots above soil. The banyan sends roots to the ground from its branches. An old banyan's root system can shelter dozens of people—above ground.

Bald cypress trees and knees growing in Florida swamp

PEOPLE AND STRANGE PLANTS

Many of these strange and curious plants are important to people. Ferns and air plants are widely used to decorate homes and yards. The shady banyan tree is planted on lawns, along streets, and on golf courses.

The bald cypress, the tree with knees, is prized for its lumber.

Seaweed is used for food in some countries. In the future, it may be even more important as food.

Glossary

bromeliad (bro MILL ee ad) — a group of tropical plants that live as epiphytes

chlorophyll (KLOR o fill) — green matter found in plants and active in making food

epiphyte (EHP uh fite) — any of several kinds of plants that grow on other plants, usually trees, and take their nourishment from the air and rain

species (SPEE sheez) — within a group of closely-related living things, one certain kind, such as a *red* mangrove tree

INDEX